Digital Traceability In Food Safety

Blockchain, IoT, and Supply Chain Transparency

Alfi Sophian

Digital Traceability In Food Safety

Blockchain, IoT, and Supply Chain Transparency

ELIVA PRESS

ISBN: 978-99993-4-007-6

Cover Design: Eliva Press
Cover Image: www.ingimage.com
Email: info@elivapress.com
Website: www.elivapress.com

This Eliva Press imprint is published by the registered company Eliva Press Global Ltd. part of Eliva Press S.R.L. Publishing Group

The registered company address is: Pope Hennessy Street Level 2, Hennessy Tower Port Louis, Mauritius
Eliva Press S.R.L. Publishing Group legal address is: Bd. Cuza-Voda 1/4 of. 21, Chisinau, Moldova, Europe

DIGITAL TRACEABILITY IN FOOD SAFETY

Blockchain, IoT, and Supply Chain Transparency

Foreword

The global food system is one of the most complex and consequential supply chains ever devised by humanity. From a single grain of wheat harvested in the fields of Kansas to a loaf of bread on a supermarket shelf in Tokyo, the journey involves hundreds of intermediaries, dozens of processing steps, and thousands of individual transactions. Yet for most of history, this journey has been invisible — opaque to consumers, difficult to audit for regulators, and nearly impossible to trace rapidly when something goes wrong.

The consequences of this opacity are severe. Every year, contaminated food sickens an estimated 600 million people worldwide, killing approximately 420,000. Foodborne illness costs the global economy over $110 billion annually in healthcare costs and lost productivity alone. When contamination events occur — from E. coli outbreaks linked to leafy greens to salmonella traced to peanut butter — the inability to quickly identify the source results in massive over-recalls, destroying the livelihoods of innocent producers and wasting enormous quantities of safe food.

This reference book arrives at a pivotal moment. Three converging technological forces — blockchain distributed ledgers, the Internet of Things (IoT), and advanced data analytics — are fundamentally reshaping what is possible in food traceability. Together, these technologies offer the prospect of near-real-time visibility across entire supply chains, tamper-resistant records of every transaction and transformation, and the ability to trace contamination from consumer to field in minutes rather than days or weeks.

This volume is designed to serve as a comprehensive guide for the full spectrum of stakeholders engaged with digital food traceability: food scientists and safety professionals seeking to understand the technological landscape; supply chain managers tasked with implementing traceability systems; IT professionals building digital infrastructure for food companies; regulators crafting policy frameworks for the new technological environment; researchers exploring the frontier of food systems innovation; and entrepreneurs developing the next generation of food safety solutions.

The book proceeds from foundational concepts through technical implementation to policy and future horizons. Each chapter builds on the previous, but each can also stand alone as a reference for specific topics. Case studies throughout the text ground abstract concepts in real-world applications, and the technical appendices provide detailed specifications for practitioners.

The digital transformation of food safety is not merely a technological change — it is a fundamental reimagining of trust in the food system. We hope this volume contributes to that transformation.

Table of Contents

Chapter 1: Food Safety in the Digital Age — An Overview 1
Chapter 2: Fundamentals of Food Traceability 12
Chapter 3: Blockchain Technology for Food Supply Chains 23
Chapter 4: Internet of Things (IoT) in Food Monitoring 34
Chapter 5: Supply Chain Transparency Frameworks 45
Chapter 6: Data Standards and Interoperability 56
Chapter 7: Implementation Strategies and Case Studies 65
Chapter 8: Regulatory Landscape and Compliance 75
Chapter 9: Challenges, Limitations, and Risk Management 84
Chapter 10: Future Directions and Emerging Technologies 92
Glossary 98
References and Further Reading 100

Chapter 1: Food Safety in the Digital Age — An Overview

The story of food safety is ultimately a story about information — who has it, who lacks it, and what happens when the right information reaches the right people at the right time. For centuries, food safety depended primarily on local knowledge: consumers bought from farmers they knew, butchers they trusted, and bakers whose practices they could observe. The industrialization and globalization of food production fundamentally disrupted this information ecosystem, creating long, anonymous supply chains in which the provenance and handling history of food products became increasingly opaque.

1.1 The Scale of the Food Safety Challenge

The dimensions of the food safety problem are staggering. According to the World Health Organization, unsafe food causes 200 different diseases ranging from diarrhea to cancer. The burden falls disproportionately on low- and middle-income countries, but no part of the world is immune. High-income countries with sophisticated regulatory systems still experience significant foodborne illness outbreaks — the United States alone experiences approximately 48 million cases of foodborne illness annually, resulting in 128,000 hospitalizations and 3,000 deaths.

The economic costs extend far beyond healthcare. Recalls triggered by contamination events can cost food companies tens or hundreds of millions of dollars. The 2008 Sanlu milk scandal in China, in which baby formula was adulterated with melamine, killed six infants and sickened 300,000 others, ultimately destroying a $3 billion company and severely damaging consumer trust in Chinese food products globally. The 2015 E. coli outbreak linked to Chipotle Mexican Grill restaurants resulted in over $500 million in lost market capitalization.

Key Statistic

According to the FDA's analysis of the 2018 romaine lettuce E. coli outbreak, it took investigators 11 days to trace the contamination source back to a specific farm in Yuma, Arizona — during which time 210 people in 36 states were sickened and a nationwide recall disrupted the entire romaine supply chain. With digital traceability systems, this process could potentially be completed in under 2 seconds.

1.2 The Evolution of Traceability Systems

Food traceability is not new. The concept of tracking food products through supply chains has existed for decades, driven by regulatory requirements and industry quality management programs. Early traceability systems were paper-based: lot numbers, batch records, shipping manifests, and receiving logs maintained in physical files. While functional in principle, these systems were slow to query, prone to human error, vulnerable to fraud, and nearly impossible to integrate across organizational boundaries.

The first generation of digital traceability systems in the 1990s and 2000s translated these paper processes into electronic databases. Enterprise Resource Planning (ERP) systems allowed large food companies to track production lots through their own operations. Barcode scanning and Electronic Data Interchange (EDI) enabled some degree of automated data capture at transaction points. However, these systems remained fundamentally siloed — each company maintained its own records, and tracing a product across multiple supply chain tiers required time-consuming manual data collection from multiple parties.

The second generation, emerging in the 2010s, introduced cloud-based platforms and more sophisticated data standards, enabling limited cross-organizational data sharing. Companies like TraceLink, Infor Nexus, and others built networks that could connect supply chain partners. However, fundamental challenges of data ownership, competitive sensitivity, and the lack of universally accepted standards limited adoption.

We are now entering what can be described as the third generation of food traceability: digital systems built on blockchain distributed ledgers, enriched with IoT sensor data, and designed from the ground up for supply chain-wide transparency. This generation is characterized by several distinguishing features that represent genuine discontinuities from previous approaches.

Traditional Traceability	Digital Traceability
Paper-based records	Blockchain distributed ledger
Company-specific databases	Industry-wide shared platforms
Manual data entry	Automated IoT data capture
Slow query times (days to weeks)	Near-instantaneous querying
Vulnerable to fraud/alteration	Tamper-evident cryptographic records
No real-time monitoring	Continuous sensor monitoring
Limited cross-company visibility	End-to-end supply chain transparency

1.3 Key Drivers of Digital Transformation

Several converging forces are accelerating the adoption of digital traceability systems in the food industry. Understanding these drivers is essential context for the technical discussions that follow.

1.3.1 Regulatory Pressure

Regulatory frameworks around the world are increasingly mandating enhanced traceability capabilities. The U.S. Food Safety Modernization Act (FSMA) of 2011, the most sweeping reform of U.S. food safety law in more than 70 years, established requirements for preventive controls and enhanced record-keeping that have driven significant technology investment. The FDA's Food Safety Traceability Rule, finalized in 2022, specifically requires companies handling high-risk foods to maintain and provide Key Data Elements (KDEs) for Critical Tracking Events (CTEs) in the supply chain, with a two-year compliance timeline.

In Europe, the European Union's Farm to Fork Strategy, a core component of the European Green Deal, explicitly calls for enhanced food traceability as part of a broader effort to build a sustainable, transparent food system. The EU's General Food Law (Regulation EC 178/2002) already requires one-step-up, one-step-back traceability for all food businesses, and proposals are being developed to extend this to more comprehensive chain-wide traceability.

1.3.2 Consumer Demand for Transparency

Consumer expectations regarding food transparency have undergone a significant shift. Surveys consistently show that substantial majorities of consumers — particularly younger generations — want to know where their food comes from, how it was produced, and under what conditions. A 2022 study by Label Insight found that 94% of consumers are more likely to be loyal to a brand that offers complete transparency, while 39% said they would switch to a new brand if they considered it more transparent.

This demand extends beyond safety to encompass sustainability, animal welfare, labor practices, and environmental impact. Digital traceability systems that can provide verified information on these dimensions are increasingly seen as competitive differentiators rather than merely compliance requirements.

1.3.3 Technology Maturation and Cost Reduction

The economics of digital traceability have shifted dramatically over the past decade. Blockchain platforms have matured from experimental technology to enterprise-ready infrastructure. The cost of IoT sensors has declined by an order of magnitude, making it economically viable to instrument cold chains, growing environments, and processing facilities that were previously cost-prohibitive. Cloud computing has eliminated the capital expense of building data infrastructure, reducing barriers to entry for smaller food companies.

1.3.4 High-Profile Contamination Events

Perhaps no single driver has been more powerful than the repeated demonstration of the costs of inadequate traceability. Each significant contamination event — and there have been dozens of major incidents globally in recent years — generates renewed urgency around the need for better systems. The Jack in the Box E. coli outbreak of 1993, the 2008 Peanut Corporation of America salmonella outbreak, the 2011 German E. coli sprout outbreak that killed 53 people — these events have served as forcing functions for industry and regulatory action.

1.4 The Digital Traceability Ecosystem

Digital traceability in food safety is not a single technology but an ecosystem of complementary technologies, standards, and practices. At the center of this ecosystem are three core technological pillars: blockchain distributed ledgers for creating tamper-evident transaction records; the Internet of Things for automated, real-time data collection; and advanced analytics for transforming raw data into actionable insights.

These technologies do not operate in isolation. They are supported by data standards frameworks like GS1 that provide the common language necessary for cross-organizational data sharing; by cloud computing infrastructure that provides scalable, accessible data storage and processing; by mobile computing that puts traceability data in the hands of workers throughout the supply chain; and by artificial intelligence systems that can identify patterns and anomalies in the vast data streams these technologies generate.

> **Definition: Digital Traceability**
>
> Digital traceability in food safety refers to the ability to track and trace the movement, transformation, and handling of food products through the supply chain using electronic systems. Complete traceability encompasses both forward tracing (tracking a product from production to consumption) and backward tracing (tracking from consumption back to production source), and should capture not only physical movement but also environmental conditions, processing events, and ownership changes.

1.5 Structure of This Reference

This reference is organized to provide both breadth and depth across the digital food traceability landscape. The following chapters address the core technical domains — traceability fundamentals (Chapter 2), blockchain (Chapter 3), IoT (Chapter 4), and supply chain transparency frameworks (Chapter 5) — before moving to enabling infrastructure (Chapter 6), real-world implementation (Chapter 7), the regulatory environment (Chapter 8), challenges and risk management (Chapter 9), and future horizons (Chapter 10).

Throughout the text, key terms are highlighted and defined; these definitions are consolidated in the Glossary at the end of the volume. Technical specifications of interest primarily to practitioners are provided in appendices to avoid interrupting the narrative flow for readers seeking conceptual understanding. Case studies drawn from real-world implementations appear throughout and are clearly distinguished from the surrounding analytical text.

Chapter 2: Fundamentals of Food Traceability

Before diving into the specific technologies that enable digital food traceability, it is essential to establish a clear conceptual foundation. What exactly is traceability? What information must be captured, by whom, and in what form? What are the critical events and data elements that give a traceability system its utility? This chapter provides that foundation, drawing on established standards frameworks and regulatory requirements to define the essential vocabulary and concepts that underpin everything that follows.

2.1 Defining Traceability in the Food Context

The Codex Alimentarius Commission — the joint FAO/WHO body that sets international food standards — defines traceability as 'the ability to follow the movement of a food through specified stage(s) of production, processing and distribution.' This definition is deliberately broad, encompassing multiple modalities of tracking and multiple stages of the supply chain. In practice, food traceability systems vary enormously in scope, granularity, and purpose.

A more operational definition distinguishes between internal and external traceability. Internal traceability refers to a company's ability to track the movement and transformation of products within its own operations — linking incoming materials to outgoing products. External traceability refers to the ability to follow products across organizational boundaries — from supplier to processor to distributor to retailer. Comprehensive chain traceability integrates both, creating an unbroken thread of information from field to fork.

2.1.1 Traceability Dimensions

A complete traceability record has three essential dimensions. The first is breadth — the extent of the supply chain covered. A system with limited breadth might trace only the last step of the supply chain (from distributor to retailer), while a comprehensive system would cover every step from primary production through consumption. The second dimension is depth — the number of steps in either direction a system can trace. One-up, one-down systems can identify the immediate source and destination of a product but cannot trace further. Full chain systems can trace from any point to all other points. The third dimension is precision — the granularity at which products are tracked. A low-precision system might track at the product category or lot level; a high-precision system might track at the individual item level.

Breadth	Depth	Precision
Limited	One step (distributor to retailer)	Product category level
Moderate	Two to three steps	Lot/batch level
Comprehensive	Full chain coverage	Individual item level

2.2 Critical Tracking Events and Key Data Elements

The FDA's Food Safety Traceability Rule, issued under FSMA Section 204, introduced two concepts that have become foundational to modern traceability system design: Critical Tracking Events (CTEs) and Key Data Elements (KDEs). These concepts provide a structured framework for determining what data must be captured and when.

2.2.1 Critical Tracking Events (CTEs)

Critical Tracking Events are defined points in the supply chain at which traceability records must be created. The FDA identifies six categories of CTEs: harvesting (the initial collection of raw agricultural commodities); cooling (the first cooling of raw agricultural commodities after harvest); initial packing (the packing of raw agricultural commodities into shipping containers); first land-based receiving (first receipt of fish or shellfish); shipping (the transfer of food products to another person); and receiving (the receipt of food products from another person). These events represent the minimum set of checkpoints at which records must be created; many organizations choose to capture data at additional points.

2.2.2 Key Data Elements (KDEs)

For each Critical Tracking Event, the FDA's rule specifies Key Data Elements that must be recorded. These data elements are designed to provide sufficient information to enable rapid tracing when a contamination event occurs. They include: Traceability Lot Codes (TLCs) that identify the specific lot of food being tracked; dates associated with the CTE; quantities; product descriptions; and location information. The specific KDEs required vary by CTE type and product category.

FDA Traceability Lot Code (TLC)

The Traceability Lot Code is the linchpin of the FDA's traceability framework. A TLC is a descriptor (either a number or alphanumeric combination) assigned to a specific lot of food at the time it is packed. The TLC allows regulators and industry to rapidly link a product in the marketplace back to its specific production lot and origin. Under the FDA rule, TLCs must be maintained at each CTE and must be accessible within 24 hours of a regulatory request.

2.3 The One-Up, One-Down Principle and Its Limitations

Many existing food safety regulations, including the EU's General Food Law, are built on the 'one-up, one-down' principle: each business in the food chain must know who supplied their food (one-up) and to whom they supplied it (one-down). This approach ensures that a contamination event can theoretically be traced by following the chain link by link.

In practice, this approach has significant limitations. The chain can only be traced as fast as each link responds to requests for information, and each link must be individually contacted and queried. In major outbreaks, this manual process can take days or weeks — time during which contaminated food continues to be consumed and people continue to get sick. The process is also vulnerable to poor record-keeping at any single link, and to intentional fraud.

Digital traceability systems address these limitations by creating a continuous, integrated record that can be queried instantaneously. Instead of calling each supplier and asking for records, investigators can query a database that already contains all the records — and that can produce a complete provenance history in seconds.

2.4 Traceability in the Farm-to-Fork Journey

To understand the scope of the traceability challenge, it is useful to trace the journey of a specific product — let us use a simple example, a bag of baby spinach from a California farm to a consumer in New York.

The journey begins at a farm, where seeds are planted, grown, and harvested. The farm itself has a complex input history: seeds from a seed supplier, water from irrigation systems, fertilizers and pesticides from agricultural suppliers, equipment maintained by service providers. At harvest, workers cut the spinach, which is loaded into field containers and transported to a cooling facility, typically within hours of harvest to maintain freshness and safety.

At the cooling facility, the spinach is washed, inspected, and packed into clamshell containers. Lot codes are assigned at this point, linking each container to a specific field, harvest date, and processing run. The containers are loaded into refrigerated trucks and transported to distribution centers — potentially making multiple stops along the way.

At distribution centers, products from many farms are consolidated and then redistributed to retail locations. A single retail pallet might contain products from a dozen different farms and processing facilities. At the retail store, the spinach is stocked on shelves and sold to consumers.

This entire journey — from field to fork — might involve 10 to 20 separate organizations, each maintaining their own records in their own format. A complete traceability system must integrate all of these records into a coherent, queryable chain.

2.5 Traceability System Architecture

A traceability system must accomplish several distinct functions: data capture (recording information about products and events), data storage (maintaining records in a durable, queryable form), data transmission (moving information between parties), data integration (combining records from multiple sources), and data access (enabling authorized users to query and receive information).

Different technological approaches take different positions on how to fulfill these functions. Traditional centralized database systems concentrate storage and integration in a single repository managed by one party. Federated systems distribute storage across multiple organizations while providing integration through standardized interfaces. Blockchain-based systems create shared, distributed ledgers that provide integration without centralized control.

2.5.1 Centralized vs. Distributed Architectures

Centralized traceability architectures offer simplicity and ease of querying but require a trusted third party to manage the central repository and create data sovereignty concerns for participating organizations. They are also single points of failure — if the central system is unavailable or compromised, the entire traceability network is affected.

Distributed architectures, including blockchain-based systems, eliminate the need for a trusted central authority and give each participant control over their own data. They are more resilient to failures and attacks. However, they are more complex to implement and query, and require strong governance frameworks to ensure consistent participation.

2.6 Traceability Metrics and Performance Standards

Evaluating the performance of a traceability system requires clear metrics. Industry and regulatory bodies have developed several key performance indicators for traceability systems that organizations can use to assess their own capabilities and identify improvement opportunities.

- Traceability breadth: percentage of supply chain tiers covered by the system
- Traceability depth: number of supply chain steps that can be traced in each direction
- Precision: granularity of product identification (item, case, pallet, lot, batch)
- Response time: time required to generate a complete provenance history in response to a query
- Completeness: percentage of transactions for which complete KDE records exist
- Accuracy: rate of errors or discrepancies in traceability records
- Integration: percentage of supply chain partners with compatible systems

The FDA has set a target response time of 24 hours for food companies to provide KDE records in response to regulatory requests. Leading companies using digital traceability systems have demonstrated the ability to produce complete chain-wide provenance histories in minutes or even seconds — a 1,000-fold improvement over manual paper-based processes.

Chapter 3: Blockchain Technology for Food Supply Chains

Blockchain technology has attracted extraordinary attention in recent years across many industries, and the claims made for it have sometimes outrun the reality. In the food safety domain, however, blockchain addresses a set of problems — the need for trusted, tamper-evident, multi-party records without a central authority — with notable precision. This chapter provides a thorough grounding in blockchain technology as it applies to food supply chains, examining both the genuine promise and the real limitations of this approach.

3.1 Blockchain Fundamentals

A blockchain is a distributed ledger — a database that is shared and synchronized across multiple computers (nodes) in a network. Unlike a conventional database managed by a single organization, a blockchain ledger is maintained simultaneously by all participants in the network, and changes to the ledger require consensus among participants according to defined rules.

The 'chain' metaphor refers to the data structure used to organize records. Transactions are grouped into blocks, and each block contains a cryptographic hash — a mathematical fingerprint — of the previous block. This chaining of blocks means that altering any historical record would change its hash, which would invalidate the hash stored in the next block, and so on down the chain. Detecting tampering thus requires only checking that the hashes in the chain are consistent — a computation that can be performed by any participant at any time.

3.1.1 Types of Blockchain Networks

For food supply chain applications, the relevant distinction is between public and permissioned (private or consortium) blockchains. Public blockchains like Bitcoin and Ethereum are open to participation by anyone, and records are visible to all. Permissioned blockchains restrict participation to a defined set of organizations and may restrict read access to authorized parties.

Public Blockchain	Permissioned Blockchain
Open to all participants	Restricted to known participants
Fully transparent records	Configurable privacy controls
Computationally intensive (PoW)	Computationally efficient (PoA/PoS)
Slow transaction throughput	High transaction throughput
No governance control	Governed by consortium rules
Examples: Bitcoin, Ethereum	Examples: Hyperledger Fabric, Quorum

For most food supply chain applications, permissioned blockchain is the appropriate choice. The participants in a food supply chain are known organizations with established legal relationships; there is no need for the open-participation model of public blockchains. Permissioned blockchains also offer significantly better performance — Hyperledger Fabric, a widely used permissioned blockchain platform, can process thousands of transactions per second, compared to the roughly 7 transactions per second achievable on the Bitcoin network.

3.2 How Blockchain Addresses Food Safety Challenges

The value of blockchain for food traceability derives from several specific properties of the technology. Understanding these properties — and, critically, their limitations — is essential for evaluating blockchain's appropriate role in traceability systems.

3.2.1 Immutability

Once a record is written to a blockchain, it cannot be altered without the alteration being detectable. This immutability property is particularly valuable in contexts where there is potential for fraud — for example, false claims about the origin of products, manipulation of food safety test results, or falsification of cold chain records. While blockchain immutability does not prevent false information from being entered in the first place (the 'garbage in, garbage out' problem that we will examine later), it does ensure that once information is recorded, it cannot be quietly changed after the fact.

3.2.2 Transparency and Auditability

All authorized participants in a permissioned blockchain can read the complete transaction history, providing a shared view of supply chain events. This eliminates the information asymmetries that characterize traditional supply chains, in which each participant has visibility only into their own slice of the chain. Regulators and auditors can review the complete record without relying on the cooperation of individual companies to provide their own records.

3.2.3 Decentralized Trust

In a blockchain network, trust is established through the consensus mechanism of the network rather than through reliance on a trusted central authority. This is particularly valuable in the food industry, where competitive dynamics make companies reluctant to share data with systems controlled by competitors or by organizations perceived as having conflicting interests. A blockchain network governed by a neutral consortium can provide the shared infrastructure that competitive parties require.

Case Study: Walmart and IBM Food Trust

In 2016, Walmart partnered with IBM to pilot a blockchain-based traceability system using the IBM Food Trust platform built on Hyperledger Fabric. The pilot dramatically illustrated blockchain's potential: tracing a package of sliced mangoes from a Walmart store in the United States back to the farm where they were grown took 6 days, 18 hours, and 26 minutes using traditional methods. Using the blockchain system, the same trace took 2.2 seconds. Following the success of this pilot, Walmart announced in 2018 that all suppliers of fresh leafy greens to its stores in the United States would be required to use the IBM Food Trust system. The initiative has since expanded to other produce categories and geographies.

3.3 Smart Contracts in Food Safety

Smart contracts are self-executing code stored on a blockchain that automatically enforce predefined rules and conditions. When specified conditions are met — for example, a shipment arriving at a warehouse — the smart contract automatically executes a defined action — recording the receipt event, triggering a payment, or generating an alert. Smart contracts are a powerful extension of basic blockchain record-keeping, enabling automated enforcement of food safety protocols.

In food safety applications, smart contracts can be used to automatically flag shipments that have violated temperature thresholds, to reject payments for products that fail quality tests, to automatically generate regulatory documentation when required conditions are met, and to enforce contractual terms related to food safety standards. This automation reduces reliance on human compliance and creates enforceable, auditable records of safety protocol adherence.

3.3.1 Smart Contract Use Cases

- Cold Chain Compliance: Automatically flagging shipments where temperature excursions are detected by IoT sensors
- Quality Gate Enforcement: Preventing products that fail laboratory tests from proceeding to the next supply chain stage
- Recall Automation: Automatically identifying and quarantining all products derived from a contaminated lot
- Supplier Certification Management: Tracking and flagging expiring food safety certifications
- Payment Automation: Releasing payments to suppliers when food safety conditions are verified
- Regulatory Reporting: Automatically generating and submitting required compliance reports

3.4 Major Blockchain Platforms for Food Safety

Several blockchain platforms have emerged as leaders in the food safety domain. Each has distinct characteristics, governance models, and technological approaches. The selection of a platform should be based on careful evaluation of these characteristics against the specific requirements of the intended application.

3.4.1 IBM Food Trust

IBM Food Trust, built on Hyperledger Fabric, is the most widely deployed blockchain platform in the food industry. It provides a permissioned network with configurable privacy controls, allowing participants to share specific data with specific counterparties while keeping other data confidential. Key features include a GS1-standard data model, integration with existing enterprise systems, and a network of thousands of participants across the food industry.

3.4.2 TE-Food

TE-Food is a global food traceability ecosystem built on its own public blockchain. It focuses on farm-to-consumer traceability and includes mobile applications for data capture at the farm level, making it suitable for traceability in developing country contexts where formal supply chain infrastructure may be limited. TE-Food has been deployed extensively in Southeast Asia and Eastern Europe.

3.4.3 Ethereum-Based Solutions

Several organizations have built food traceability applications on the Ethereum blockchain or on private blockchains that implement the Ethereum Virtual Machine (EVM). These solutions benefit from the rich ecosystem of Ethereum development tools and smart contract standards, though the throughput limitations of the public Ethereum network typically necessitate the use of private or layer-2 deployments for production food safety applications.

3.5 The 'Oracle Problem' and Data Integrity

One of the most important limitations of blockchain for food safety is the 'oracle problem': blockchain can guarantee the integrity of data once it is entered on the chain, but it cannot guarantee the accuracy of data at the point of entry. If a worker scans a batch code incorrectly, or deliberately enters false information, the blockchain will faithfully record and preserve that inaccurate data.

This limitation means that blockchain is necessary but not sufficient for food safety traceability. It must be combined with robust physical verification systems — including IoT sensors, camera systems, and human oversight — that minimize the opportunity for inaccurate or fraudulent data entry. The integration of IoT sensors with blockchain systems, discussed in the next chapter, is a key strategy for addressing the oracle problem by automating data capture and reducing reliance on manual data entry.

The Oracle Problem in Practice

Consider a shipment of fresh fish that is claimed to have been maintained at 4°C throughout transport. A blockchain record stating that the temperature was maintained is only as trustworthy as the process by which that record was created. If the temperature record was entered manually by a driver with incentives to falsify it, the blockchain record is unreliable. If the record was generated automatically by a calibrated IoT temperature logger that transmitted data directly to the blockchain via a secure wireless connection, the record is much more trustworthy. IoT integration is the primary solution to the oracle problem.

3.6 Governance Frameworks for Food Blockchain Networks

The technical properties of blockchain are only as valuable as the governance frameworks that surround them. A blockchain network requires clear rules about who can participate, what data must be recorded, what privacy protections apply, how disputes are resolved, and how the network evolves over time. Poorly governed blockchain networks have failed in numerous industries; successful food blockchain initiatives are distinguished by strong governance frameworks developed through industry collaboration.

The GS1 standards organization has played a critical role in developing governance frameworks for food blockchain networks, providing the data standards and network governance models that enable interoperability across the industry. Major food industry blockchain initiatives — including those led by Walmart, Carrefour, Nestlé, and others — have generally adopted GS1 standards as the basis for data representation.

Chapter 4: Internet of Things (IoT) in Food Monitoring

If blockchain provides the ledger for recording food supply chain events, the Internet of Things provides the sensors that generate the data those records contain. IoT technology enables automated, real-time monitoring of food products throughout the supply chain — from field conditions during growing, to temperature during transport, to storage conditions in warehouses and retail environments. This chapter examines the IoT technologies used in food safety monitoring, their capabilities and limitations, and how they integrate with blockchain and other traceability systems.

4.1 IoT Architecture for Food Supply Chains

An IoT system for food supply chain monitoring consists of several layers: sensing devices that capture physical data; connectivity infrastructure that transmits data; edge computing systems that process data locally; cloud platforms that store and analyze data; and application interfaces that present data to users and integrate with other systems. Understanding each layer is important for designing effective systems.

4.1.1 Sensing Devices

The foundation of any IoT food monitoring system is the sensing device — the hardware that directly measures conditions in the physical environment. A wide variety of sensor types are used in food safety applications, measuring parameters from temperature and humidity to specific chemical compounds and biological agents.

Sensor Type	Parameter Measured	Application
Temperature sensor	Product and environment temp.	Cold chain, storage monitoring
Humidity sensor	Relative humidity	Grain storage, produce warehousing
Gas sensor (CO_2, O_2, ethylene)	Atmosphere composition	Modified atmosphere packaging
pH sensor	Acidity level	Fermentation monitoring
RFID tag	Product identity	Item-level tracking
GPS tracker	Location	Transport route monitoring
Camera/imaging	Visual condition	Quality inspection, contamination detection
Biosensor	Pathogen presence	Rapid food safety testing

4.2 Temperature Monitoring in Cold Chains

Temperature control is the most critical single parameter in food safety for many product categories. Cold chain failures — breaks in the continuous refrigeration required from production to consumption — are responsible for a significant proportion of foodborne illness cases and food waste globally. The World Health Organization estimates that 10% of the global disease burden of foodborne illness could be eliminated through improved cold chain management.

Traditional cold chain monitoring relied on periodic manual temperature checks or simple data loggers that recorded temperatures at fixed intervals and had to be physically retrieved for data download. Modern IoT temperature monitoring systems provide continuous, real-time temperature data transmitted wirelessly to cloud platforms, with automated alerting when thresholds are exceeded.

4.2.1 Cold Chain Monitoring Technologies

Several distinct technologies are used for cold chain temperature monitoring, each with different characteristics regarding cost, accuracy, communication range, and power requirements.

RFID-based temperature loggers combine radio frequency identification for product identification with integrated temperature sensors. They can be scanned at checkpoints to retrieve accumulated temperature records without requiring a continuous wireless connection. Near Field Communication (NFC) loggers are a variant that allow temperature records to be read by smartphones, enabling consumers to verify cold chain compliance at the point of purchase.

Bluetooth Low Energy (BLE) temperature loggers provide continuous temperature recording with periodic automatic transmission to nearby gateways or smartphones. They are well-suited for monitoring conditions within refrigerated vehicles, warehouses, and retail display cases. Their range is limited to approximately 100 meters, but this is typically sufficient for their intended applications.

Cellular-connected temperature loggers provide continuous, long-range temperature reporting without relying on local gateway infrastructure. They are particularly valuable for monitoring temperature in transit on vehicles traveling routes where local WiFi or Bluetooth infrastructure is absent. The ongoing data costs associated with cellular connectivity are higher than those for short-range technologies.

Global Cold Chain Losses

According to the Food and Agriculture Organization of the United Nations, approximately one-third of all food produced globally for human consumption is wasted. In developing countries, much of this loss occurs at the production and distribution stages due to inadequate cold chain infrastructure. In high-income countries, losses are more often at the retail and consumer stages. IoT-enabled cold chain monitoring has been demonstrated to reduce losses by 5-10% in pilot programs, representing a significant economic and sustainability benefit in addition to the food safety implications.

4.3 RFID and Barcode Technologies

Radio Frequency Identification (RFID) and barcode technologies are the primary means by which individual products and cases are identified and tracked throughout the supply chain. While less 'glamorous' than newer sensing technologies, these identification systems are the backbone of food traceability — without accurate product identification, even the best environmental monitoring data cannot be linked to specific products.

4.3.1 Barcode Systems

Barcodes have been the standard product identification technology since their introduction in the early 1970s. GS1-standard barcodes — including the ubiquitous EAN/UPC barcode on retail products, the GS1-128 barcode used on shipping containers, and the GS1 DataMatrix used for compact, high-information encoding — encode product identifiers, lot numbers, expiration dates, and other data in a machine-readable format.

Two-dimensional barcodes, including QR codes and GS1 DataMatrix, can encode significantly more data than one-dimensional barcodes and can be read with standard smartphone cameras, making them suitable for consumer-facing applications. QR codes on food packaging are increasingly being used to provide consumers with access to detailed provenance information stored on traceability platforms.

4.3.2 RFID Systems

RFID technology encodes information in electronic tags that can be read by radio wave interrogation without requiring line-of-sight scanning. This allows RFID readers to scan entire pallets or truckloads of products simultaneously, dramatically accelerating the data capture process at receiving and shipping points. Ultra-High Frequency (UHF) RFID, operating at 860-960 MHz, is the dominant technology for supply chain applications, with reading ranges of several meters.

The primary barrier to universal RFID adoption in food supply chains has historically been tag cost — while barcode labels cost fractions of a cent, RFID tags have historically cost 10-20 cents, making item-level tagging economically impractical for low-cost food products. However, tag costs have declined significantly and continue to fall, and case- and pallet-level RFID tagging is now economically viable for most food supply chain applications.

4.4 Smart Packaging and Consumer-Facing IoT

An emerging frontier in food IoT is smart packaging — packaging that incorporates sensing and communication capabilities to provide real-time information about product condition and provenance. Smart packaging technologies range from simple time-temperature indicators (TTIs) that provide a visual indication of cold chain integrity, to sophisticated electronic systems that continuously monitor and report product condition.

Consumer-facing smart packaging typically uses Near Field Communication (NFC) or QR code interfaces to allow consumers to access product information using their smartphones. Applications include accessing detailed provenance information ('scan to learn where your salmon was caught'), verifying cold chain compliance ('this product maintained throughout transport'), and checking authenticity ('this product is genuine').

4.5 Environmental Monitoring in Food Production

IoT monitoring is not limited to the distribution phase of the food supply chain. Precision agriculture technologies use IoT sensors to monitor growing conditions, enabling farmers to optimize inputs and identify potential contamination risks at the source. Soil moisture sensors, weather stations, water quality monitors, and aerial imaging systems all contribute to a comprehensive picture of the growing environment.

In food processing facilities, IoT sensors monitor sanitation conditions, equipment performance, and environmental parameters that could affect food safety. Air quality sensors can detect airborne pathogens; surface cleanliness sensors verify sanitation effectiveness; equipment monitoring systems identify mechanical failures before they cause food safety incidents.

4.6 Edge Computing and Real-Time Processing

The volume of data generated by IoT food monitoring systems can be enormous — a large food processing facility might have thousands of sensors generating data points every few seconds. Transmitting all of this data to a central cloud platform for processing would create significant bandwidth and latency challenges. Edge computing addresses this by processing data locally, at or near the point of generation, and transmitting only relevant events and summaries to the cloud.

In food safety applications, edge computing enables real-time response to safety-critical events without waiting for data to travel to a cloud platform and back. A temperature excursion detected by an edge computing system can trigger an immediate local alert within milliseconds, while the centralized cloud platform provides long-term storage, analytics, and integration with traceability systems.

Chapter 5: Supply Chain Transparency Frameworks

Technology alone does not create supply chain transparency. Technologies must be deployed within organizational and institutional frameworks that define who participates, what information is shared, who has access to what data, and how the system is governed. This chapter examines the frameworks — both conceptual and practical — that enable supply chain transparency in the food industry, drawing on industry standards, regulatory requirements, and best practices from successful implementations.

5.1 The Transparency Spectrum

Supply chain transparency exists on a spectrum from complete opacity — in which no supply chain information is publicly available — to radical transparency — in which all supply chain data is freely accessible to all stakeholders. In practice, food supply chain transparency systems operate somewhere in the middle of this spectrum, with different levels of access for different stakeholders.

A useful framework distinguishes three levels of supply chain transparency. Operational transparency provides full visibility to direct supply chain partners — a manufacturer knows everything about their direct suppliers, and their direct customers know everything about them, but neither can see further up or down the chain. Extended transparency provides visibility to all supply chain tiers within a company's supply network, including multiple levels of sub-suppliers. Radical transparency extends visibility to consumers and the public, allowing anyone to access information about a product's provenance and journey.

Transparency Level	Accessible To
Operational Transparency	Direct supply chain partners only
Extended Transparency	All tiers of supply network
Radical Transparency	Consumers and public

5.2 GS1 Standards: The Language of Supply Chain Transparency

GS1 is the global standards organization that develops and maintains the most widely used supply chain data standards in the world. GS1 standards provide the common language that enables supply chain participants to exchange product information accurately and efficiently. For food traceability systems, GS1 standards are essentially the foundation on which all practical implementations must be built.

5.2.1 GS1 Identification Standards

GS1 identification standards assign globally unique identifiers to physical entities in the supply chain. The Global Trade Item Number (GTIN) identifies a specific product type. The Global Location Number (GLN) identifies a physical location (a warehouse, a farm, a processing facility). The Serial Shipping Container Code (SSCC) identifies a specific logistics unit (a pallet, a case). The Batch/Lot Number, combined with a GTIN, identifies a specific production lot. Together, these identifiers allow any product, at any level of the supply chain hierarchy, to be uniquely identified and linked to relevant records.

5.2.2 GS1 Data Capture Standards

GS1 data capture standards define how identification information is encoded in machine-readable form — barcodes, RFID tags, and digital formats. The GS1 Application Identifier (AI) system provides a standardized format for encoding multiple data elements in a single barcode or RFID tag, including GTIN, lot number, expiration date, and serial number.

5.2.3 GS1 Data Sharing Standards: EPCIS

The Electronic Product Code Information Services (EPCIS) standard is GS1's framework for sharing supply chain event data. EPCIS defines a common data model and interface for recording and querying supply chain events — including what product was involved (the 'what'), where the event occurred (the 'where'), when it occurred (the 'when'), and why it occurred (the 'why'). EPCIS version 2.0, released in 2022, includes enhanced support for IoT-generated data and blockchain integration, making it the preferred standard for next-generation traceability systems.

EPCIS and the FDA's Traceability Rule

The FDA's Food Safety Traceability Rule is designed to be technology-neutral — it specifies what data must be captured (KDEs) and at what points (CTEs), but does not mandate specific technology approaches. However, the GS1 EPCIS standard provides a natural mapping for the KDEs and CTEs required by the rule, and several major food traceability platforms have implemented the rule's requirements using EPCIS as the underlying data model. Organizations that have already deployed EPCIS-based traceability systems will find compliance with the FDA rule significantly simplified.

5.3 The Layered Transparency Model

A practical framework for implementing supply chain transparency recognizes that different stakeholders have different information needs and different levels of trust. A layered model provides different levels of information access to different stakeholder groups, balancing the desire for transparency with legitimate concerns about competitive sensitivity and data privacy.

At the foundation is the full traceability record — the complete set of KDEs for all CTEs, accessible to regulators and authorized internal users with a demonstrated need for full chain visibility. This layer includes commercially sensitive information such as supplier identities, pricing, and volumes. The second layer provides extended traceability information — provenance data including origin, certifications, and environmental conditions — shared with direct supply chain partners on a need-to-know basis. The third layer is consumer-facing, providing the information that consumers can access through product packaging, QR codes, or retailer websites.

5.4 Multi-Stakeholder Governance Models

Effective supply chain transparency requires governance structures that bring together the multiple parties whose cooperation is required. No single company — not even the largest retailer or food manufacturer — controls all the parties in its supply chain. Creating a traceability network requires voluntary participation by hundreds or thousands of independent businesses, each of which must see a sufficient business case for participation.

Successful multi-stakeholder governance models for food traceability have typically involved a combination of regulatory mandates (creating a floor of required transparency), retailer requirements (leveraging the power of major retailers to require supplier participation), industry associations (providing neutral governance structures for cross-competitor collaboration), and technology platform providers (offering the shared infrastructure that enables participation).

5.4.1 Retailer-Led Models

Major retailers have been among the most powerful drivers of supply chain transparency, using their purchasing power to require suppliers to participate in traceability programs. Walmart's requirement that leafy greens suppliers join the IBM Food Trust blockchain network is the most prominent example, but similar requirements have been implemented by Carrefour, Ahold Delhaize, and other major retailers globally. Retailer-led models are effective at driving rapid adoption within a specific retail channel but can create fragmentation if different retailers require participation in incompatible systems.

5.4.2 Industry Consortium Models

Industry consortium models bring together multiple companies — including competitors — to develop shared traceability infrastructure. The Global Dialogue on Seafood Traceability (GDST), which developed the GDST 1.1 standard for seafood traceability, is an example of a successful industry consortium that produced a widely adopted interoperability framework. Consortium models are better suited to achieving industry-wide standards but require significant investment in governance and are slower to develop than retailer-led initiatives.

5.5 Consumer-Facing Transparency Applications

The ultimate beneficiaries of food supply chain transparency are consumers — the people who eat the food. Consumer-facing transparency applications make supply chain data accessible to individual consumers, typically through smartphone interfaces and QR codes on product packaging.

The range of information that can be made available to consumers through these applications is vast: the specific farm where produce was grown; the vessel on which fish were caught; the certifications held by processors; the environmental footprint of production; the social and labor conditions throughout the supply chain. Different consumer segments show different levels of interest in different types of information, and the design of consumer-facing applications must balance comprehensiveness with usability.

Chapter 6: Data Standards and Interoperability

The potential of digital food traceability can only be realized if systems deployed by different organizations across the supply chain can communicate with each other. This interoperability challenge — enabling data created in one system to be read, understood, and acted upon by a different system — is one of the central challenges of food traceability implementation. Data standards provide the common language that makes interoperability possible.

6.1 The Interoperability Challenge

The food supply chain involves hundreds of thousands of companies, ranging from individual farmers with simple smartphones to multinational corporations with sophisticated ERP systems. These companies use different software platforms, different data formats, and different data management practices. A traceability system that works perfectly within the confines of a single large company's ERP system is of limited value if it cannot exchange data with the small suppliers and regional distributors that constitute the majority of most companies' supply chains.

Interoperability challenges exist at multiple levels. Syntactic interoperability requires that data be formatted in ways that receiving systems can parse — using agreed file formats, message structures, and communication protocols. Semantic interoperability requires that the meaning of data elements be shared — that when one system sends a 'lot number,' the receiving system understands it to be the same construct as its own 'batch code.' Process interoperability requires that the business processes implied by data exchange be understood and compatible — that the 'shipping event' recorded by a supplier corresponds to the 'receiving event' expected by the customer.

6.2 Core Standards for Food Traceability

Several categories of standards are relevant to food traceability interoperability. These standards exist at different levels of the standards hierarchy and serve different functions.

6.2.1 Identification Standards

Identification standards establish globally unique identifiers for products, locations, and organizations, ensuring that all parties in the supply chain can refer to the same entities using the same codes. GS1 identification standards (GTIN, GLN, SSCC) provide the primary identification framework for food supply chains globally, and their use is essentially mandatory for companies seeking to exchange data with major retailers and distributors.

6.2.2 Data Format Standards

Data format standards define how information is encoded and transmitted between systems. JSON (JavaScript Object Notation) and XML (Extensible Markup Language) are the dominant generic data formats for modern supply chain data exchange. GS1's EPCIS standard defines a specific data model and message format for supply chain events, built on XML and JSON. EDI (Electronic Data Interchange) standards, including the widely used ANSI X12 and EDIFACT standards, define formats for business documents like purchase orders, advance ship notices, and invoices that carry traceability-relevant information.

6.2.3 Communication Protocol Standards

Communication protocol standards define how data is exchanged between systems. RESTful APIs (Application Programming Interfaces) using HTTP have become the dominant method for real-time data exchange in modern traceability systems. EPCIS defines a standardized REST API for querying supply chain event data. For IoT devices, MQTT (Message Queuing Telemetry Transport) is the dominant protocol for transmitting sensor data due to its low bandwidth requirements and support for unreliable networks.

> **EPCIS 2.0: The New Standard for Food Traceability**
>
> GS1's EPCIS 2.0, released in 2022, represents a significant advance over its predecessor. Key enhancements include native support for IoT sensor data (including temperature, humidity, and other physical parameters) through the new Sensor Element; improved support for linked data and semantic interoperability through alignment with JSON-LD; enhanced security and privacy features; and a RESTful API that simplifies integration with modern software platforms. EPCIS 2.0 is the recommended standard for new food traceability system implementations and provides a natural implementation framework for the FDA's Food Safety Traceability Rule.

6.3 Master Data Management

Traceability event data — records of shipments, receipts, and transformations — is only meaningful in the context of master data that defines the entities being tracked. Master data includes product definitions (what is a GTIN?), location definitions (where is GLN 0614141123452?), and party definitions (who is the company with Global Company Prefix 0614141?). Without accurate, synchronized master data, traceability event records cannot be correctly interpreted.

Master data management (MDM) for food traceability involves ensuring that all participants in a traceability network are working with consistent, accurate, and up-to-date master data. The GS1 Global Data Synchronization Network (GDSN) provides a cloud-based infrastructure for synchronizing master data across supply chain partners, and many food retailers require suppliers to maintain their product data in the GDSN as a condition of doing business.

6.4 Interoperability Between Blockchain Networks

A significant challenge for food blockchain traceability is the multiplicity of competing platforms. Walmart requires suppliers to use IBM Food Trust; a competing retailer might require a different platform. A supplier selling to both retailers must participate in two incompatible blockchain networks, duplicating data entry efforts and creating reconciliation challenges.

Several approaches are being developed to address cross-platform interoperability. The GS1 EPCIS standard, by providing a common data model for supply chain events, creates a potential common language that could bridge different blockchain implementations. The Global Dialogue on Seafood Traceability (GDST) has developed an interoperability standard for the seafood sector that enables data exchange between different traceability platforms. Industry-wide adoption of common data standards is the most sustainable long-term solution to the fragmentation problem.

6.5 Data Quality Management

The value of a traceability system is directly proportional to the quality of the data it contains. Poor data quality — including missing records, incorrect identifiers, inconsistent formatting, and duplicate entries — undermines traceability capability at every level. Data quality management is therefore a critical, if often unglamorous, component of food traceability system implementation.

Key data quality dimensions for traceability systems include completeness (are all required KDEs present?), accuracy (do records reflect actual events?), timeliness (are records created when required?), consistency (are the same entities described using the same identifiers throughout?), and linkage (can records be successfully joined across organizational boundaries?). Each of these dimensions requires specific management processes and technical controls.

- Completeness: Automated validation checks that reject submissions missing required data fields
- Accuracy: IoT automation to replace manual data entry where possible; exception alerting for anomalous data values
- Timeliness: Workflow integration that embeds data capture in operational processes rather than treating it as a separate step
- Consistency: Master data management systems that enforce consistent use of identifiers
- Linkage: Matching algorithms that reconcile shipping records from suppliers with receiving records from customers

Chapter 7: Implementation Strategies and Case Studies

The previous chapters have established the conceptual and technical foundations of digital food traceability. This chapter turns to the practical challenge of implementation: how do organizations actually deploy these technologies in real-world food supply chains? Drawing on documented case studies and implementation experience from across the food industry, this chapter provides a framework for planning and executing traceability technology projects.

7.1 Implementation Readiness Assessment

Before embarking on a digital traceability implementation, organizations need to honestly assess their current state of readiness. A gap between the current state and the target state defines the scope of the implementation project. Key readiness dimensions include data infrastructure, process maturity, partner readiness, and governance capability.

7.1.1 Current State Assessment Framework

A structured assessment of current traceability capabilities should examine several dimensions. Data infrastructure readiness addresses the question of whether current systems generate, capture, and store the data needed for traceability. Are lot numbers assigned consistently? Are receiving and shipping records electronic or paper-based? Can current systems generate the KDEs required by regulations?

Process maturity examines whether current operational processes support accurate data capture. Are workers trained to scan products at the right points? Are process controls in place to prevent scanning errors? Are exception handling procedures defined for cases where scanning is not possible?

Partner readiness is often the most critical and most difficult dimension. A traceability system is only as strong as its weakest link; if major suppliers cannot provide digital traceability records, the chain is broken. An honest assessment of supplier readiness — including their technological capabilities and their willingness to participate — is essential before committing to a traceability implementation.

> **Case Study: Nestlé Open Farm Initiative**
>
> In 2018, Nestlé launched an initiative to provide consumers with transparent access to supply chain information for its Mousline potato purée brand in France, using an IBM Food Trust blockchain system. The initiative demonstrated several key implementation lessons: consumer adoption of QR code scanning was lower than anticipated, suggesting that consumer-facing transparency requires significant marketing investment to realize its potential; supply chain data quality was a more significant challenge than the blockchain technology itself; and the project required substantial education and support for small-scale potato farmers who had never previously used digital systems.

7.2 Technology Selection Framework

The selection of technology platforms for food traceability involves evaluating candidates against multiple criteria. The relative importance of these criteria will vary depending on the organization's specific situation, but a comprehensive evaluation should address the following considerations.

1. Regulatory compliance: Does the platform support capture and retrieval of all KDEs required by applicable regulations?
2. Supply chain coverage: What proportion of the relevant supply chain is already on the platform, or can be onboarded?
3. Integration capability: How well does the platform integrate with existing ERP and warehouse management systems?
4. Scalability: Can the platform handle the volume of transactions generated by the supply chain?
5. Cost: What is the total cost of ownership, including implementation, licensing, and ongoing operation?
6. Standards compliance: Does the platform implement GS1 and other relevant industry standards?
7. Vendor stability: What is the financial stability and long-term viability of the platform vendor?
8. User experience: How easy is the platform to use for workers throughout the supply chain, including those with limited technical skills?

7.3 Phased Implementation Approach

Given the complexity and scale of food supply chain traceability implementations, a phased approach is almost always appropriate. Attempting to implement comprehensive chain-wide traceability in a single project is rarely successful; the complexity is too great, the number of parties involved is too large, and the potential for disruptive failure is too high.

A proven phased approach begins with an internal traceability foundation — ensuring that the organization's own operations have robust digital traceability before extending the system to supply chain partners. This phase focuses on ensuring that lots are consistently identified, that ERP systems generate accurate event records, and that workers are trained and processes are in place for accurate data capture.

The second phase extends traceability to direct suppliers — the first tier of the upstream supply chain. This phase typically involves the most significant supplier onboarding challenge, as it requires working with suppliers of varying sizes and technological capabilities to ensure they can provide digital traceability records in a compatible format.

Subsequent phases extend coverage progressively further up and down the supply chain, and add additional data dimensions — environmental sensor data, quality test results, certification records — to the basic event records established in earlier phases.

7.4 Major Case Studies

7.4.1 Walmart Leafy Greens Initiative

Walmart's requirement that suppliers of fresh leafy greens to its U.S. stores use the IBM Food Trust blockchain traceability system is the most widely cited case study in food blockchain traceability. The initiative was announced in September 2018, following a series of E. coli outbreaks linked to romaine lettuce that affected consumers across the United States and Canada.

The implementation involved onboarding more than 100 suppliers, ranging from large national growers to small regional farms. IBM provided a portal-based data entry interface for smaller suppliers who lacked the technical capability for direct system integration. Walmart's scale — it is the largest food retailer in the United States — gave it the leverage to require supplier participation as a condition of continuing to sell through its stores, demonstrating the power of retailer mandates in driving supply chain-wide adoption.

Post-implementation evaluations confirmed a dramatic reduction in trace time: the 11-day trace that had been required using traditional methods could be completed in approximately 2.2 seconds using the blockchain system. The initiative has subsequently been extended to additional produce categories and has served as a model for similar programs by other retailers.

7.4.2 Thailand Shrimp Traceability Program

Thailand is the world's largest exporter of shrimp, but the Thai shrimp industry faced significant reputational damage in the early 2010s following investigations that uncovered labor abuses and illegal fishing practices in the supply chain. In response, a multi-stakeholder initiative involving the Thai government, major shrimp processors, and international retailers implemented a comprehensive digital traceability system using GS1 standards and RFID technology.

The system tracks shrimp from fishing vessels or aquaculture farms through processing plants to export. Each batch of shrimp is assigned a GS1-standard lot code at the point of landing, and this code follows the product through all subsequent processing steps. The system has been credited with improving labor practices and reducing illegal fishing by making it easier to identify supply chain violations.

7.4.3 China's National Food Safety Traceability Platform

China operates one of the world's largest national food traceability systems, the National Food Safety Traceability Platform (NFSTP), managed by the State Administration for Market Regulation. The platform covers a wide range of food categories, including infant formula, pork, vegetables, and dairy products, and requires participating businesses to register product information and report key supply chain events.

The Chinese experience illustrates both the potential and the challenges of government-mandated traceability systems. The scale of coverage achievable through regulatory mandate is difficult to replicate through voluntary industry initiatives. However, enforcement challenges in a vast and diverse food industry have limited data quality in some segments of the system.

7.5 Change Management and Training

Perhaps the most commonly underestimated dimension of traceability implementation is change management — the human side of the technology deployment. New traceability systems require workers throughout the supply chain to change their behaviors: scanning products at different points, entering data into new systems, following new protocols for exception handling.

Successful implementations invest heavily in training — not just in how to use the new systems, but in why traceability matters and how the new system contributes to food safety and business objectives. Workers who understand the purpose of traceability are more likely to comply with new procedures and to report exceptions rather than working around them.

Chapter 8: Regulatory Landscape and Compliance

The regulatory environment for food traceability is evolving rapidly, driven by the convergence of technological capability, consumer demand, and the lessons of major food safety incidents. This chapter surveys the current regulatory landscape in major markets, examines how digital traceability systems support compliance, and provides guidance on navigating the complex multi-jurisdictional regulatory environment faced by companies operating in global food supply chains.

8.1 United States Regulatory Framework

The regulatory framework for food traceability in the United States is primarily established through the Food Safety Modernization Act (FSMA) of 2011 and the regulations that implement it. FSMA represented a fundamental shift in U.S. food safety philosophy — from a reactive, response-to-illness approach to a proactive, prevention-based approach.

8.1.1 FSMA Section 204: Enhanced Traceability

Section 204 of FSMA directed the FDA to establish additional recordkeeping requirements for the foods most associated with foodborne illness outbreaks. The resulting Food Safety Traceability Rule, finalized in November 2022 and effective from January 2023, establishes requirements for entities that manufacture, process, pack, or hold foods on the Food Traceability List (FTL). The FTL includes several categories of fresh produce, cheeses, shell eggs, nut butters, ready-to-eat deli salads, and certain seafood products.

The rule requires covered entities to establish and maintain records of Key Data Elements (KDEs) for Critical Tracking Events (CTEs), to assign and maintain Traceability Lot Codes, and to provide records to the FDA within 24 hours of request. The rule is technology-neutral — it specifies what data must be kept but not how it must be kept — meaning that paper records can in principle satisfy the requirements, though digital systems are far more efficient.

8.1.2 Other FSMA Regulations with Traceability Implications

Several other FSMA regulations have implications for traceability, even if they do not specifically require traceability systems. The Preventive Controls rules for Human Food and Animal Food require hazard analysis and risk-based preventive controls, with associated documentation that creates a de facto traceability record of key process parameters. The Foreign Supplier Verification Program (FSVP) requires importers to verify that foreign suppliers are producing food in a manner that meets U.S. food safety standards, which requires supply chain visibility into foreign suppliers' operations.

> **FDA FSMA Section 204 Compliance Timeline**
> The FDA's Food Safety Traceability Rule established a phased compliance timeline. Large businesses (annual sales over $1 million) were required to comply by January 20, 2026. Small businesses (annual sales $1 million or less) and very small businesses have extended compliance dates. The FDA has stated its intention to provide enforcement discretion during an initial period while the industry adapts to the new requirements. Companies that have invested in digital traceability systems are well-positioned to demonstrate compliance.

8.2 European Union Regulatory Framework

The European Union has a comprehensive food safety regulatory framework that has been in place since the early 2000s, anchored by the General Food Law (Regulation EC 178/2002). The EU framework predates many of the digital technologies examined in this book, but it has proven adaptable to the new technological environment.

8.2.1 General Food Law Traceability Requirements

Article 18 of the General Food Law requires all food and feed businesses in the EU to be able to identify the food and feed suppliers from which they have received products (one step back) and the businesses to which they have supplied products (one step forward). This one-up, one-down traceability requirement has been in force since January 2005 and applies to all food businesses regardless of size.

8.2.2 Farm to Fork Strategy

The European Green Deal, launched in 2019, includes the Farm to Fork Strategy as a key component of the EU's plan for a sustainable food system. The Farm to Fork Strategy explicitly calls for enhanced traceability across the food chain, including digital labeling systems that would provide consumers with comprehensive information about food products' sustainability, nutritional, and provenance characteristics. While the specific regulatory requirements flowing from the Farm to Fork Strategy are still being developed, the direction of travel is clearly toward significantly enhanced traceability requirements.

8.3 Asian Regulatory Frameworks

Asia is home to some of the world's most sophisticated food traceability regulatory systems, as well as some of the most significant food safety challenges. China, Japan, and South Korea have all invested heavily in food traceability regulation, driven by major domestic food safety incidents and by the need to reassure consumers about the safety of the food supply.

8.3.1 China

China's food safety regulatory framework has been substantially strengthened following the 2008 melamine infant formula scandal and subsequent incidents. The Food Safety Law, revised in 2015, established requirements for traceability systems for key food categories. Specific regulations require traceability for infant formula, dairy products, pork, vegetables, and seafood.

8.3.2 Japan

Japan has implemented sector-specific food traceability requirements for beef (following BSE scares), rice (following the 2008 rice adulteration scandal), and seafood. The beef traceability system, implemented in 2003, assigns individual ID numbers to all cattle born in Japan and tracks each animal through slaughter, processing, and retail sale. Japanese consumers can access the traceability record for any domestic beef product by entering the ID number printed on the product label.

8.4 Regulatory Compliance Management

For companies operating in multiple markets, managing compliance with a complex and evolving multi-jurisdictional regulatory environment is a significant challenge. Digital traceability systems that implement recognized international standards (GS1, EPCIS) provide the best foundation for multi-jurisdictional compliance, as these standards have been explicitly recognized or implicitly accepted by regulatory frameworks in most major markets.

Key practices for regulatory compliance management include maintaining a regulatory monitoring program to track developments in all relevant jurisdictions; engaging with regulatory agencies proactively to understand evolving expectations; participating in industry associations that engage with regulators on traceability policy; and documenting the compliance rationale for all traceability system design decisions.

8.5 Documentation and Audit Readiness

Beyond meeting specific regulatory requirements, food companies must be able to demonstrate their compliance in the event of an audit or investigation. Digital traceability systems provide a significant advantage in this regard: the ability to quickly retrieve and present complete traceability records, in response to regulatory requests, is both a regulatory requirement and a practical business necessity.

Audit readiness for food traceability requires more than having the technical capability to retrieve records. It also requires documentation of the traceability system design, including which CTEs are captured, which KDEs are recorded at each CTE, how records are stored and protected, and how the system handles exceptions. This documentation, combined with regular internal audits and drills simulating regulatory queries, provides the foundation for confident regulatory engagement.

Chapter 9: Challenges, Limitations, and Risk Management

No technology is a silver bullet, and the honest assessment of digital food traceability systems must include a clear-eyed examination of their limitations and the challenges organizations face in deploying them. This chapter provides that examination, identifying the key technical, organizational, and systemic challenges that must be managed for successful traceability implementation, and outlining risk management approaches for each.

9.1 Technical Challenges

9.1.1 Connectivity and Infrastructure Gaps

Many segments of the food supply chain — remote farms, developing country supply chains, vessels at sea, cold storage facilities with limited wireless infrastructure — face connectivity challenges that complicate IoT deployment and real-time data capture. These gaps are particularly acute at the beginning of supply chains, where the most important provenance information is generated but technological infrastructure is often most limited.

Approaches to connectivity challenges include the use of satellite IoT networks (which can provide coverage in remote areas where cellular networks are absent), store-and-forward approaches (where data is collected locally and transmitted when connectivity is available), and hybrid paper/digital systems for the least connected segments of supply chains.

9.1.2 Scalability and Performance

Food supply chains generate enormous volumes of transactions. A major food retailer might receive hundreds of thousands of product deliveries per week, each generating multiple traceability events. Blockchain systems, in particular, must be carefully architected for performance at these scales — not all blockchain platforms are capable of handling production food supply chain volumes without significant engineering investment.

9.1.3 Data Privacy and Security

Supply chain data can be highly commercially sensitive. Supplier identities, pricing, volumes, and production capacities are competitive intelligence that companies are legitimately reluctant to share broadly. Blockchain systems that make all data visible to all participants create real data privacy risks that can deter participation.

Privacy-preserving techniques — including selective disclosure (sharing specific data elements with specific parties without revealing the full record), zero-knowledge proofs (cryptographic techniques that can verify claims about data without revealing the underlying data), and tokenization (replacing sensitive identifiers with non-sensitive placeholders in shared records) — can address these concerns but add complexity to system design.

Data Security Considerations

Food traceability systems are increasingly targets for cyberattacks, as adversaries recognize that disrupting traceability infrastructure can cause significant economic damage during food safety incidents. Key security considerations include access control and authentication (ensuring that only authorized parties can read or write traceability records), data encryption (protecting data in transit and at rest), audit logging (maintaining records of all access to traceability data), and resilience planning (ensuring that traceability systems remain functional during cyberattacks or other disruptions to IT infrastructure).

9.2 Organizational Challenges

9.2.1 Supply Chain Fragmentation

Food supply chains are highly fragmented, involving many small and medium-sized enterprises (SMEs) with limited technological capabilities. Onboarding thousands of small farmers, processors, and distributors onto sophisticated digital traceability systems is a formidable challenge. Solutions must be designed with the least sophisticated participants in mind — mobile-first, offline-capable, multilingual, and requiring minimal training.

9.2.2 Incentive Alignment

Traceability creates costs for supply chain participants who must invest in new technologies and processes. The benefits of traceability, however, are often asymmetric — accruing primarily to parties further down the supply chain (retailers and consumers) while the costs are borne by parties further up (farmers and processors). Creating incentive structures that fairly distribute the costs and benefits of traceability is a critical governance challenge.

9.2.3 Organizational Culture and Resistance to Change

Perhaps the most persistent challenge in food traceability implementation is organizational culture. Workers and managers throughout the supply chain may resist new traceability requirements, perceiving them as administrative burden without clear benefit to their day-to-day work. Overcoming this resistance requires clear communication of the business case for traceability, visible leadership commitment, and training that emphasizes the food safety rationale for the requirements.

9.3 The 'Garbage In, Garbage Out' Problem

The fundamental data integrity challenge for any traceability system is the 'garbage in, garbage out' problem: a traceability system is only as trustworthy as the accuracy of the data that enters it. Blockchain immutability guarantees that data, once entered, cannot be quietly altered — but it cannot guarantee that the original entry was accurate.

Sources of inaccurate data in traceability systems include honest errors (workers scanning the wrong barcode, entering incorrect quantities), process failures (records created for systems rather than for actual products), and deliberate fraud (falsifying provenance records, mislabeling products). Each of these sources requires different mitigation approaches.

9.3.1 Mitigation Strategies

- Automate data capture wherever possible using IoT sensors and readers to reduce reliance on manual entry

- Implement validation checks that flag data entries inconsistent with expected values (quantities outside expected ranges, locations inconsistent with shipping routes)
- Use cross-validation between records from different parties (supplier shipping records should match customer receiving records)
- Implement physical verification systems (weight checks, visual inspection, sampling) that create independent verification of key parameters
- Design processes so that accurate data capture is the path of least resistance, rather than an additional burden on workers
- Conduct regular data quality audits and provide feedback to supply chain partners on data quality issues

9.4 Fraud and Adulteration Risks

Food fraud — the intentional substitution, dilution, or mislabeling of food products for economic gain — is a multi-billion dollar problem that directly undermines food safety and consumer trust. Digital traceability systems can both deter and detect certain types of food fraud, but they are not a complete solution.

Digital traceability creates a more complete audit trail that increases the risk of detection for fraudulent substitution. If a consignment of premium extra virgin olive oil is substituted with cheap sunflower oil at some point in the supply chain, a detailed traceability record that shows the product moving from an olive oil producer through the chain creates a clear framework for investigating the fraud. However, traceability records can themselves be falsified — a fraudulent actor can create false records as well as substituting products. The combination of traceability records with physical verification (DNA testing, stable isotope analysis, and other analytical techniques) provides the most robust fraud deterrence.

9.5 Risk Management Framework

A structured risk management framework for digital food traceability should identify and address risks across technical, organizational, data quality, and security dimensions. The framework should include risk identification (systematically identifying potential failure modes), risk assessment (evaluating the likelihood and impact of each failure mode), risk mitigation (implementing controls to reduce likelihood or impact), and risk monitoring (continuously assessing the effectiveness of controls).

Chapter 10: Future Directions and Emerging Technologies

The technologies and practices examined in this reference represent the state of the art as of the mid-2020s. The field is evolving rapidly, however, and several emerging technologies promise to further transform digital food traceability over the coming decade. This final chapter examines these emerging technologies, discusses the future trajectory of food traceability, and offers a perspective on the path to a fully transparent, fully traceable global food system.

10.1 Artificial Intelligence and Machine Learning

Artificial intelligence and machine learning are beginning to play significant roles in food traceability, primarily through their ability to extract insights from the large volumes of data that traceability systems generate. AI applications in food traceability span several domains.

10.1.1 Predictive Risk Analytics

AI models can analyze traceability data to identify patterns that predict food safety risks before they manifest as consumer illness. Environmental sensor data, processing records, supplier quality histories, and external data sources (weather patterns, disease outbreaks in growing regions) can be combined in predictive models that flag high-risk lots before they enter distribution. Early detection of potential contamination events can prevent recalls rather than merely enabling faster responses to them.

10.1.2 Anomaly Detection

Machine learning algorithms trained on normal supply chain patterns can identify anomalies that may indicate data quality problems or food fraud. An unusually long transit time, a temperature deviation too brief to appear in manual review but visible in high-frequency sensor data, or a quantity discrepancy between shipping and receiving records — these patterns, invisible to human reviewers of large datasets, can be automatically detected by machine learning systems.

10.1.3 Computer Vision for Quality Control

Computer vision systems powered by deep learning are achieving human-level accuracy in visual quality inspection tasks for many food products. Automated inspection systems that can identify defects, contamination, and quality issues in real time can both improve food safety and generate traceability records that document the quality status of products at each inspection point.

10.2 Digital Twins in Food Supply Chains

A digital twin is a virtual model of a physical system that is continuously updated with real-time data from the physical system and can be used for simulation, optimization, and prediction. Digital twins are being developed for food supply chains, creating virtual representations of entire supply chains that can be used to simulate the consequences of disruptions, optimize logistics decisions, and identify potential food safety risks.

A food supply chain digital twin might incorporate real-time shipment location data, temperature sensor readings, weather forecasts along transport routes, production schedules at processing facilities, and demand signals from retailers — creating a comprehensive, continuously updated model of the state of the supply chain. This model can be used to proactively reroute shipments threatened by temperature excursions, to identify potential recall impact before a contamination event is confirmed, and to optimize inventory positioning to minimize food waste.

10.3 Advanced Sensor Technologies

The sensing technologies available for food monitoring are advancing rapidly, with new sensors capable of detecting contaminants and quality parameters that were previously measurable only with expensive laboratory equipment.

10.3.1 Biosensors for Rapid Pathogen Detection

Traditional microbiological testing for foodborne pathogens requires 24-72 hours for results — too slow for practical in-line food safety screening. Next-generation biosensors based on technologies including surface acoustic wave (SAW) devices, aptamer-based sensors, and microfluidic lab-on-a-chip systems are achieving detection of common food pathogens in minutes rather than days, with sensitivity approaching that of laboratory culture methods.

10.3.2 Spectroscopic and Hyperspectral Imaging

Near-infrared (NIR) and Raman spectroscopy can identify the chemical composition of food products non-destructively, enabling rapid authentication (confirming that a product is what it claims to be) and quality assessment. Hyperspectral imaging systems that capture spectroscopic data across the entire surface of a food product can identify contamination, bruising, and quality defects that are invisible to conventional cameras.

10.4 Web3 and Decentralized Traceability

The emerging Web3 paradigm — a vision of the internet built on decentralized protocols, blockchain infrastructure, and user-controlled data — has potential applications in food traceability. Smart contract-based traceability systems that operate without central intermediaries, tokenized supply chain incentives that reward accurate data provision, and self-sovereign supply chain data that supply chain participants own and control rather than sharing with platform providers are all areas of active development.

Non-fungible tokens (NFTs) have been explored as a mechanism for certifying the provenance and authenticity of premium food products — creating a unique digital artifact tied to a physical product that can travel through the supply chain alongside the product and be verified by consumers at the point of purchase. While the technology is immature and the use cases are still being developed, the combination of NFTs with physical anti-counterfeiting measures (smart labels, DNA markers) could provide a powerful authenticity verification system for high-value food products.

10.5 The Path to Universal Food Traceability

Looking further ahead, the trajectory of food traceability technology points toward a future in which every food product is fully traceable from production to consumption, in real time, with full transparency to all authorized stakeholders. Reaching this vision requires progress on several fronts simultaneously.

Technological progress must continue on cost reduction (making IoT sensors and connectivity affordable for the smallest farms and processors), on automation (reducing the burden of data capture on workers), and on interoperability (ensuring that systems deployed in different parts of the supply chain can exchange data seamlessly). Regulatory progress must create clear, harmonized requirements that create a level playing field and ensure that the benefits of traceability are distributed fairly across the supply chain. Industry progress must address governance challenges, creating inclusive multi-stakeholder structures that serve the interests of all supply chain participants.

The 2-Second Trace Vision

The Food Safety Tech industry has articulated a vision of '2-second traceability': the ability to trace any food product from fork to farm in under 2 seconds in the event of a food safety incident. The Walmart-IBM Food Trust pilot demonstrated that this is technically achievable today — at least for supply chains that have been fully digitized. Achieving it universally, across all food categories and all supply chain configurations, will require a decade or more of continued technological deployment, regulatory evolution, and industry collaboration. But the destination is now in sight.

10.6 Sustainability and Digital Traceability

Food traceability and food system sustainability are deeply connected. The same data infrastructure that supports food safety also supports sustainability measurement and reporting — tracking the carbon footprint of products from field to consumer, verifying claims about sustainable fishing and farming practices, and documenting compliance with deforestation-free sourcing commitments.

Emerging regulatory frameworks in Europe and elsewhere are creating requirements for supply chain due diligence on sustainability issues that parallel the requirements for food safety traceability. The EU's Corporate Sustainability Reporting Directive (CSRD) and the proposed Supply Chain Due Diligence Act are examples of regulatory requirements that will drive significant investment in supply chain visibility for sustainability purposes. Companies that have already invested in digital food safety traceability are well positioned to extend these systems to meet emerging sustainability transparency requirements.

10.7 Conclusion

The digitization of food traceability represents one of the most significant transformations in food safety in the modern era. The convergence of blockchain, IoT, and advanced data analytics is creating the possibility of a food system in which contamination events can be identified and contained in minutes rather than days, in which consumers have unprecedented visibility into the provenance and journey of their food, and in which the trust that was once provided by local knowledge can be reconstructed at global scale through technology.

Realizing this possibility requires more than technology investment. It requires organizational change, regulatory evolution, governance innovation, and a sustained commitment across the food industry to the shared goal of a safe, transparent, and sustainable food system. The technologies described in this reference provide the tools; the work of building the food system we need remains fundamentally a human endeavor.

Glossary

Application Programming Interface (API)

A software interface that allows different systems to communicate with each other by exchanging data and commands according to defined specifications.

Batch/Lot

A quantity of food produced under the same conditions and identified by a common code. Lot traceability links all units produced in a single production run to a shared origin and processing history.

Blockchain

A distributed ledger technology in which transactions are recorded in blocks that are chained together using cryptographic hashes, creating an immutable record that is maintained across multiple nodes in a network.

Cold Chain

The uninterrupted sequence of refrigerated production, storage, and distribution activities, along with associated equipment and logistics, that maintain the quality and safety of temperature-sensitive foods.

Critical Tracking Event (CTE)

A point in the food supply chain at which traceability records must be created, as defined by the FDA's Food Safety Traceability Rule. CTEs include harvesting, cooling, initial packing, shipping, and receiving.

Digital Twin

A virtual model of a physical system that is continuously updated with real-time data and can be used for simulation, optimization, and prediction.

EPCIS (Electronic Product Code Information Services)

A GS1 standard that defines a data model and interface for sharing supply chain event data, addressing the what, where, when, and why of supply chain events.

Food Safety Modernization Act (FSMA)

U.S. legislation enacted in 2011 that shifted the focus of U.S. food safety regulation from responding to contamination to preventing it, establishing requirements for preventive controls, enhanced record-keeping, and traceability.

GS1

The global standards organization that develops and maintains the most widely used supply chain data standards, including the Global Trade Item Number (GTIN), Global Location Number (GLN), and EPCIS.

Hyperledger Fabric

An open-source permissioned blockchain platform developed by the Linux Foundation, widely used for food supply chain traceability applications including IBM Food Trust.

Internet of Things (IoT)

The network of physical devices, vehicles, appliances, and other items embedded with sensors, software, and connectivity that enables them to collect and exchange data.

Key Data Element (KDE)

Specific pieces of information that must be recorded at Critical Tracking Events under the FDA's Food Safety Traceability Rule, including Traceability Lot Codes, dates, quantities, and location information.

Oracle Problem

The challenge of ensuring that data recorded on a blockchain accurately reflects physical reality. Blockchain can guarantee the integrity of data once recorded but cannot verify its accuracy at the point of entry.

Permissioned Blockchain

A blockchain network in which participation is restricted to authorized organizations, as opposed to public blockchains that are open to anyone. Permissioned blockchains are generally preferred for food supply chain applications.

RFID (Radio Frequency Identification)

A technology that uses radio waves to communicate between a reader and a tag attached to a product, enabling non-contact identification and tracking without requiring line-of-sight scanning.

Smart Contract

Self-executing code stored on a blockchain that automatically enforces predefined rules and conditions when specified events occur.

Traceability

The ability to follow the movement of a food through specified stages of production, processing, and distribution, both forward (from origin to consumption) and backward (from consumption to origin).

Traceability Lot Code (TLC)

A unique identifier assigned to a specific lot of food at the time it is packed, used as the key linking identifier in the FDA's food traceability framework.

Zero-Knowledge Proof

A cryptographic technique that allows one party to prove to another that they know a piece of information without revealing the information itself, enabling privacy-preserving verification in blockchain systems.

References and Further Reading

Regulatory Documents and Standards

- U.S. Food and Drug Administration. (2022). Food Safety Traceability Rule (21 CFR Part 1, Subpart S). Federal Register, Vol. 87, No. 225.
- European Commission. (2002). Regulation (EC) No 178/2002 of the European Parliament and of the Council — General Food Law.
- GS1. (2022). EPCIS Standard, Version 2.0. GS1 Document: GS1-EPCIS-Standard.
- Codex Alimentarius Commission. (2006). Principles for Traceability/Product Tracing as a Tool within a Food Inspection and Certification System. CAC/GL 60-2006.
- GS1. (2020). GS1 Traceability Standard, Version 2.0.
- Global Dialogue on Seafood Traceability. (2020). GDST 1.1: Seafood Traceability Technical Standard.

Books and Academic References

- Bosona, T., & Gebresenbet, G. (2013). Food traceability as an integral part of logistics management in food and agricultural supply chain. Food Control, 33(1), 32-48.
- Charlebois, S., Sterling, B., Haratifar, S., & Naing, S. K. (2014). Comparison of global food traceability regulations and requirements. Comprehensive Reviews in Food Science and Food Safety, 13(5), 1104-1123.
- Galvez, J. F., Mejuto, J. C., & Simal-Gandara, J. (2018). Future challenges on the use of blockchain for food traceability analysis. TrAC Trends in Analytical Chemistry, 107, 222-232.
- Ge, L., Brewster, C., Spek, J., Smeenk, A., & Top, J. (2017). Blockchain for agriculture and food: Findings from the pilot study. Wageningen Economic Research.
- Khanna, V. K., & Srivastava, M. (2012). Food traceability: Review of current state and challenges for the future. Comprehensive Reviews in Food Science and Food Safety, 11(2), 169-178.
- Lin, Q., Wang, H., Pei, X., & Wang, J. (2019). Food safety traceability system based on blockchain and EPCIS. IEEE Access, 7, 20698-20707.
- Sapiun, Z., Sophian, A., Abinawanto, M., Kamba, V., Damiti, S. A., & Luawo, H. (2020). Optimization of Mcfarland Turbidity standards value in determining template DNA as reference in Salmonella Typhimurium ATCC 14028 test using Real-Time PCR (QPCR). *PalArch's Journal of Archaeology of Egypt/Egyptology*, *17*(6), 10916-10922.
- Sophian, A. (2022). Escherichia coli Bacteria Test on Polluted Meatballs with Several Variations of Positive Control Concentration. *Journal Biology Science Education*, *3*(1), 32-38.
- Sophian, A. (2025). *Ancaman Parasitik: Patogenitas Toxoplasma gondii pada Manusia dan Hewan.*

- Sophian, A. (2025). *Cereulide Toxin sebagai Faktor Utama Sindrom Muntah Bacillus cereus.*
- Sophian, A. (2025). *Mencegah Bahaya Patogenitas Salmonella demi Keamanan Pangan yang Lebih Kuat.*
- Sophian, A. (2025). *Patogenitas Bacillus cereus: Peran Toksin dalam Foodborne Illness.*
- Sophian, A. (2025). *Patogenitas Cyclospora cayetanensis: Tantangan Keamanan pada Konsumsi Sayur dan Buah Segar.*
- Sophian, A. L. F. I., Purwaningsih, R. A. T. N. A., Igirisa, E. P. J., Amirullah, M. L., Lukita, B. L., & Fitri, R. A. (2021). Detection of Salmonella typhimurium ATCC 14028 and Listeria monocytogenes ATCC 7644 in processed meat products using Real-Time PCR Multiplex Method. *Asian Journal of Natural Product Biochemistry, 19*(1), 17-20.
- Sophian, A. L. F. I., Purwaningsih, R. A. T. N. A., Lukita, B. L., & Ningsih, E. C. (2020). Detection of Salmonella typhimurium ATCC 14028 in supplement health product liquid preparation using Real-Time PCR (qPCR). *Biofarmasi Journal of Natural Product Biochemistry, 18*(2), 65-69.
- Sophian, A., & Muindar, M. (2021). PENGUJIAN SALMONELLA TYPHIMURIUM ATCC 14028 PADA PRODUK SOSIS, NUGGET, BAKSO, OTAK-OTAK, TEMPURA DAN CILOK MENGGUNAKAN KIT RAPID TEST. *Journal of Experimental and Clinical Pharmacy (JECP), 1*(1).
- Sophian, A., & Purwaningsih, R. (2022). Detection of Lysteria monocytogenes in Frozen Meatballs Using Real-Time PCR. *Indonesian Food Science and Technology Journal, 6*(1), 27-30.
- Sophian, A., Purwaningsih, R., Muindar, M., Igirisa, E. P. J., & Amirullah, M. L. (2021). Use of direct PCR technique without DNA extraction in confirmation test for Salmonella typhimurium bacteria on meatball samples. *Borneo Journal of Pharmacy, 4*(4), 324-332.
- Sophian, A., Purwaningsih, R., Muindar, M., Igirisa, E. P. J., & Amirullah, M. L. (2021). Detection of Salmonella typhimurium ATCC 14028 in powder prepared traditional medicines using Real-Time PCR. *Borneo Journal of Pharmacy, 4*(3), 178-183.
- Sophian, A., Purwaningsih, R., Sutrisno, M. T., Purwadi, P., & Wahyudi, A. (2022). Detection of salmonella typhimurium bacteria on bakery products samples using boiling isolation technique. *Jurnal Farmasi Sains Dan Praktis*, 274-280.
- Sophian, A., Utaminingsih, S., Bhakti, T. S., & Rahmawati, D. (2023). Real-time PCR application in confirmation test of Salmonella Typhimurium on instant noodle. *Ho Chi Minh City Open University Journal of Science-Engineering and Technology*, 3-9.
- Tian, F. (2017). A supply chain traceability system for food safety based on HACCP, blockchain & internet of things. 2017 International Conference on Service Systems and Service Management.
- Verdouw, C. N., Wolfert, J., Beulens, A. J. M., & Rialland, A. (2016). Virtualization of food supply chains with the internet of things. Journal of Food Engineering, 176, 128-136.
- World Health Organization. (2015). WHO estimates of the global burden of foodborne diseases. WHO Press.
- Zhang, X., Qian, Z., Xu, C., Sun, D. W., & Pu, H. (2022). A comprehensive review of IoT-based food quality monitoring: Sensors, data, and intelligence. Comprehensive Reviews in Food Science and Food Safety, 21(3), 2419-2450.

Industry Reports and White Papers

- Deloitte. (2019). Blockchain: Opportunities for Health Care. Deloitte Insights.
- Food and Agriculture Organization of the United Nations. (2019). Moving Forward on Food Loss and Waste Reduction. FAO.
- IBM Institute for Business Value. (2018). IBM Food Trust: Building a Smarter Food Ecosystem. IBM Corporation.
- McKinsey & Company. (2020). COVID-19 and food safety: Rebuilding consumer trust. McKinsey Global Institute.
- PricewaterhouseCoopers. (2019). Food Safety in the Age of Digital Transformation. PwC.
- World Economic Forum. (2020). Innovation with a Purpose: Improving Traceability in Food Value Chains through Technology Innovations. WEF.
- GS1. (2023). How Digital Traceability Can Strengthen Food Safety: Aligning with FDA's New Food Safety Traceability Rule. GS1 US.

Online Resources

- FDA Food Safety Traceability Hub: www.fda.gov/food/food-safety-modernization-act-fsma/food-traceability
- GS1 Traceability Standards: www.gs1.org/standards/traceability
- IBM Food Trust: www.ibm.com/food
- Global Dialogue on Seafood Traceability: www.seafoodtraceability.org
- Hyperledger Foundation: www.hyperledger.org
- EU Farm to Fork Strategy: food.ec.europa.eu/horizontal-topics/farm-fork-strategy_en

Publisher: Eliva Press Global Ltd

Email: info@elivapress.com

www.ingramcontent.com/pod-product-compliance
Lightning Source LLC
LaVergne TN
LVHW022324200926
848481LV00004B/1414
* 9 7 8 9 9 9 3 4 0 0 7 6 *